The Fungi Kingdom

by

Farin DeBose

Copyright ©2014 Farin DeBose

Illumination Publications, Inc.

All rights reserved.

The Fungi Kingdom   by Farin DeBose

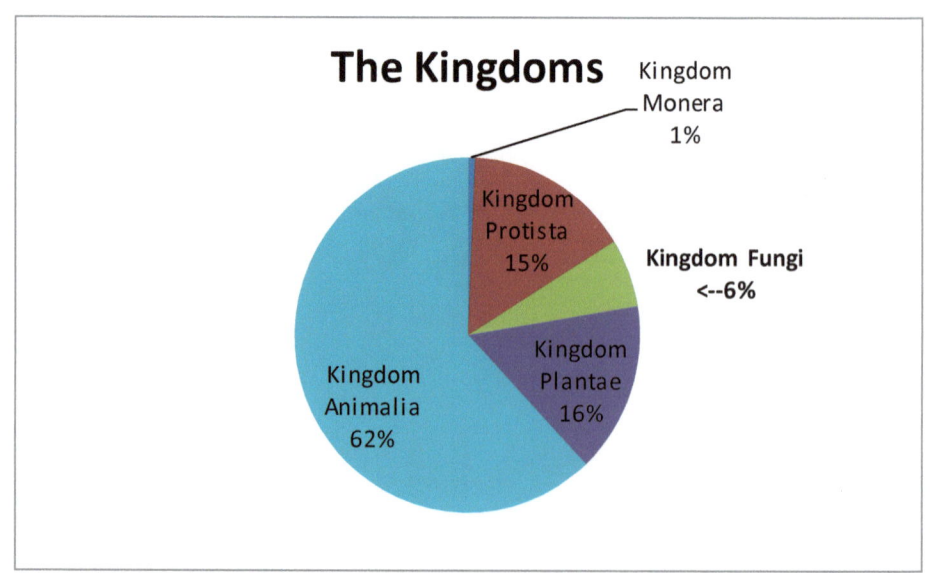

Long ago, in the beginning of time, there were five kingdoms. The two main or most popular kingdoms were the animal kingdom and the plant kingdom. But, there was another less well-known kingdom whose king wanted that to change and have greater notoriety.

The Fungi Kingdom — by Farin DeBose

The mushroom king wanted the respect of the other kings. He was rejected by the plants because he did not make his own food. Likewise, he was rejected by the animals even though he fed off living and rotted plants.

The Fungi Kingdom				by Farin DeBose

So, the mushroom king decided to establish the Kingdom of Fungi for outcasts like himself. His kingdom would welcome blights growing on crops and destroying them and also parasites spreading diseases in animals.

The Fungi Kingdom   by Farin DeBose

King Agaric and Queen Bonnet could not understand why the other kingdoms considered themselves better than the fungi. There were at least 100,000 species or different kinds of fungus!

The Fungi Kingdomby Farin DeBose

So, even though they wanted their status as laughingstocks to change, the king and queen were happy together in the Kingdom of Fungi.

The Fungi Kingdom
by Farin DeBose

One day the king's mother Chanterelle changed all that when she remarked, **"Why have you and the queen not made any spores. You have no heir."**

Queen Bonnet was upset over Chanterelle's prying because she knew that the king would listen to his mother's advice.

The Fungi Kingdom   by Farin DeBose

The King was deeply affected by his mother's words. He thought to himself, fungi reproduce by making spores:

"Fungi produce spores in reproductive structures called fruiting bodies. They produce asexually in a process called budding where fungi are genetically identical to the parent. They can also reproduce sexually when two fungi grow from joined hyphae where the spores are genetically different from either parent."

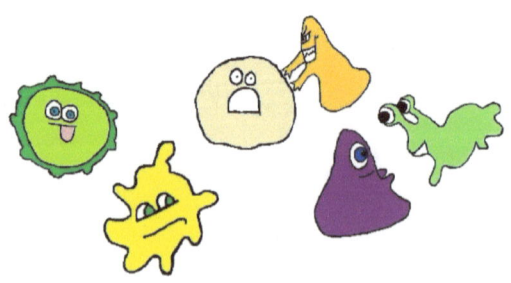

The Fungi Kingdom   by Farin DeBose

Queen Bonnet noticed the king was deep in thought and decided to bring entertainment to the palace to take his mind off of things. He needed a distraction.

The town fool came and told lots of jokes but the king hated his act. The town fool foolishly asked, **"Hey, what does a king do when he can't have any children? Forget about it!"**

The Fungi Kingdom — by Farin DeBose

The fool also laughed heartily at his own jokes. **"I'll be here all week if you need me. But, seriously folks... forget about it. It ain't gonna happen!"**

As the fool left the palace, not realizing he had bombed, he exclaimed, **"Peace out."**

The Fungi Kingdom by Farin DeBose

The queen thought, that didn't go so well. "**How about some dinner**?" she asked.

The king and his queen went to the dark damp royal dining room where algae brought in a scrumptious dead fly for each of them. The algae thought it was gross that the pair, like most fungi, fed off of dead things and rotting plants.

The Fungi Kingdom by Farin DeBose

The queen exclaimed, "A fly for dinner is a nice change and quite a feast!"

The king and his wife broke down the chemicals of the dead organism. The fly's decomposed body returned nutrients to the soil for the fungi to absorb.

The hyphae also helped them absorb water from the soil to quench their thirst after their feast.

The Fungi Kingdom
by Farin DeBose

"Do you feel better honey?" The queen noticed her king seemed better. The king's mind seemed clear and no longer focused on spores or the Fungi Kingdom's standing in the world.

King Agaric managed to "fool" his wife. He did not like being made fun of by the town fool or being ridiculed by his mother for not having offspring. He put on music and a show for his queen. They listened to the soothing song lyrics to forget about their troubles.

The Fungi Kingdom by Farin DeBose

Right on beat, they started dancing and doing the fungi rap:
**Use spores to reproduce
Your fruiting body needs life
You can do it alone
Or bud with your wife
Make sac fungi, club fungi,
or zygote fungi!**

The Fungi Kingdom by Farin DeBose

Queen Bonnet smiled and stood next to King Agaric. Maybe it was the tribal song but the conditions became right. Mushroom spores grew inside the thin walls of their gills on the underside of their umbrella-shaped caps.

The Fungi Kingdom by Farin DeBose

Soon new fungi grew from the spores. First one, then two...

The Fungi Kingdom   by Farin DeBose

It was a miracle. The king and queen finally had their wish of an heir; lots and lots of heirs.

The Fungi Kingdom   by Farin DeBose

Their offspring grew up to become decomposers and recyclers. They provided food for people. They caused disease but also fought disease—especially their daughter Princess Penicillium.

The Fungi Kingdom   by Farin DeBose

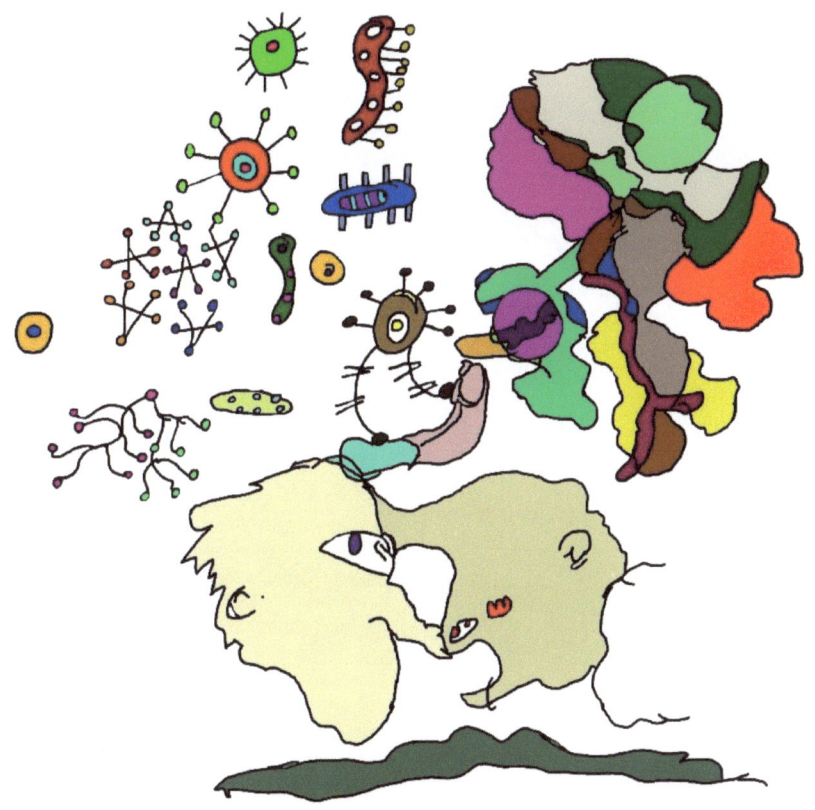

They also lived in symbiosis with other organisms, like Lichen, who lived in a mutualistic relationship with algae and autotrophic bacteria.

The Fungi Kingdom   by Farin DeBose

Most importantly, the species of the Fungi Kingdom earned the respect of the other kingdoms, most notably the Plant and Animal kingdoms.

Finally, content, King Agaric, Queen Bonnet, their family, and subjects lived happily ever after.

The End

The Fungi Kingdom by Farin DeBose

About the Book:

The Fungi Kingdom is an elementary and middle school science primer that introduces children, age 9 and up, to biology and the world around us. Biologists today have classified and divided all living things into five groups they call Kingdoms. These Kingdoms are based on how living things are the same and how they are different. The five Kingdoms presently accepted by a majority of scientists are the Monera Kingdom, the Protist Kingdom, the Fungi Kingdom, the Plant Kingdom, and the Animal Kingdom. This book focuses on the Fungi Kingdom.

About the Author:

Farin DeBose is the intelligent, creative, and unscripted author of this book for children. At age 12, Farin became a young prodigy, publishing her first children's book, *Pogo's Big Mistake*. As a young author, Farin wrote several funny tales to teach children about important topics. She has been said to have a knack for writing, demonstrating academic skill and ability but also a gift in the arts.

Farin's many interests include playing guitar, writing songs, singing, and performing. She was an avid competitive gymnast and dancer. Farin has completed honors, advance placement, and international baccalaureate courses in high school, before pursuing a Bioengineering degree at Temple University in Philadelphia.

Check out *Pogo's Big Mistake* and look for other new titles coming from this rising writer of children's educational books, including *Every Frog Praises Its Own Pond* and *Why Sloths Are So Slow*.

The Fungi Kingdom by Farin DeBose

Word Glossary

Absorb – take in or soak up. Example: sponge.

Agaric – a fungus resembling an ordinary mushroom, with a flattened cap and gills underneath.

Algae – a simple, nonflowering plant. Example: seaweed.

Animal – a living organism that feed on organic matter. They have senses, organs, and a nervous system. Example: lions.

Asexual – having no evident sex (gender) or sex organs.

Autotrophic bacteria – an organism capable of producing its own food from inorganic substances, using light or chemical energy. Example: algae, green plants, bacteria.

Blight – a plant disease caused by fungi. Example: mildew and rust.

Bonnet – mushroom with a burgundy drop bonnet and whitish gills underneath.

Budding – outgrowths or buds on the bodies of mature organisms, capable of developing into a new organism.

The Fungi Kingdom   by Farin DeBose

Club fungi – common name of species with branched, club-shaped sporophores.

Decompose – to make or become rotten; to decay or cause to decay.

Fruiting Bodies – the spore-producing organ of a fungus, like a mushroom or toadstool.

Fungi – different single-celled organisms that live by decomposing and taking in organic material in which they grow. Example: mushrooms.

Genetic – arising from or relating to a common origin.

Hyphae – branching elements that make up a fungus.

Kingdom – the highest level in which organisms are grouped. Example: Fungi Kingdom, Animal Kingdom, and Plant Kingdom.

Lichen - The mutualistic symbiotic association of a fungus with an alga or a cyanobacterium, or both, forming a crustlike or branching growth on rocks or tree trunks.

Mutualistic Relationship – when two organisms of different species work together, each benefiting from the relationship.

Nutrients – nourishment for growth and maintenance of life. Example: food that has protein, vitamins, and minerals.

The Fungi Kingdom   by Farin DeBose

Offspring – an animal's or plant's young; the product or result of something. Example: child, cub, seedling.

Organism – an individual animal, plant, or single-celled lifeform. Examples: plant, lion, fungus.

Parasite – an organism that lives in another organism, taking nutrients from its host. Example: fungus.

Penicillium – a blue mold that produces antibiotics naturally.

Plant – a living organism that takes in water and inorganic substances through its roots. Often, they are green. Examples: trees, shrubs, and grass.

Recyclers – decomposers that recycle nutrients in dead organisms and
their waste.

Sac fungi – fungus of molds and truffles.

Species – organisms capable of exchanging genes or interbreeding; not capable of breeding with members of another species.

Spores – a tiny, one-celled body produced especially by fungi.

The Fungi Kingdom by Farin DeBose

Symbiosis – a mutually beneficial relationship between different organisms living together.

Zygote fungi – resistant spherical spores living in terrestrial habitats in soil or decaying plants or animal material.

www.ingramcontent.com/pod-product-compliance
Lightning Source LLC
Chambersburg PA
CBHW041121180526
45172CB00001B/360